AF228905

BLIZZARDS

by Joanne Mattern

BrightPoint Press

San Diego, CA

© 2025 BrightPoint Press
an imprint of ReferencePoint Press, Inc.
Printed in the United States

For more information, contact:
BrightPoint Press
PO Box 27779
San Diego, CA 92198
www.BrightPointPress.com

LIBRARY OF CONGRESS CATALOGING-IN-PUBLICATION DATA

Names: Mattern, Joanne, 1963- author.
Title: Blizzards / by Joanne Mattern.
Description: San Diego, CA: BrightPoint Press, [2025] | Series: Understanding natural disasters | Includes bibliographical references and index. | Audience: Grades 7-9
Identifiers: LCCN 2024004143 (print) | LCCN 2024004144 (eBook) | ISBN 9781678208660 (hardcover) | ISBN 9781678208677 (eBook)
Subjects: LCSH: Blizzards--Juvenile literature. | Winter storms--Juvenile literature.
Classification: LCC QC926.37 .M3844 2025 (print) | LCC QC926.37 (eBook) | DDC 363.34/925--dc23/eng/20240216
LC record available at https://lccn.loc.gov/2024004143
LC eBook record available at https://lccn.loc.gov/2024004144

CONTENTS

AT A GLANCE

- Blizzards are severe winter storms that can cause great damage and loss of life.

- A winter storm becomes a blizzard when it meets three conditions. These conditions are high winds, low visibility, and a duration of at least 3 hours.

- Blizzards can form when areas of low and high pressure meet and clouds release a great deal of moisture in cold weather.

- Blizzards usually form in cold, northern parts of the world. They can also occur at high elevations.

- People in blizzards can suffer frostbite, hypothermia, and even death.

- Blizzards can cause transportation delays as snow and ice make it difficult to drive.

- Heavy snow can cause roofs to collapse. Snow and wind can bring down power lines.

- The National Weather Service issues watches, advisories, and warnings to alert people to dangerous storms.

- Planning ahead, finding safe shelter, and having plenty of emergency supplies are important ways to stay safe in a blizzard.

A STORM TO REMEMBER

It was March 1888. People on the East Coast of the United States were looking forward to spring. But as people went to bed on March 11, it started to snow. By the next morning, 10 inches (25 cm) of snow had fallen. Snow fell for the next 2 days. Up to 4 feet (1 m) of snow covered cities and towns along the East Coast. Almost all business halted. The New York Stock Exchange closed on March 13. The storm

Blizzards can bring activity in cities to a halt.

became known as the Great Blizzard
of 1888.

The blizzard hit the small towns around
Princeton, New Jersey, very hard. One man

Snowdrifts are piles of snow created by blowing wind.

who had been a child during the storm talked about it years later. The night of the storm, he was sick with the measles. The next day, he was amazed by what he saw outside. He said, "I remember looking out the window. There was a great big heap [of snow], higher than a two-and-a-half-story house, right in front of our house [. . .]. And I know that looked to me like a mountain."[1]

The snow fell until March 14. In the days following, the East Coast began to thaw. The snow melted, causing floods. The New York Stock Exchange reopened on March 15. The blizzard had caused more than $20 million of property damage. It is remembered as one of the worst blizzards in US history.

People should think twice about stepping outside during a blizzard.

THE POWER OF BLIZZARDS

The Great Blizzard of 1888 is an example of the power of wind and snow. Blizzards are a natural disaster that people should take seriously. Many parts of the world experience these frightening weather events. They're most likely to happen in the northern parts of the world, where snow is common. But they can also occur in warmer areas where snow is less common.

Blizzards make it hard to see far ahead. This is because fast winds blow snow through the air. Sometimes snow falls without causing a blizzard. But when weather conditions are right, blizzards can strike.

CHAPTER ONE

WHAT ARE BLIZZARDS?

Blizzards are snowstorms to the extreme. An Iowa newspaper first printed the word *blizzard* in this sense in the 1870s. By the 1880s, the word had spread across the country.

The National Weather Service says that a storm has to meet three conditions to count as a blizzard. First, the storm needs to have winds of at least 35 miles per hour (56 kmh). Snow and high winds combine

to limit visibility. If it's hard to see less than a quarter mile (0.4 km) away, the storm can be called a blizzard. Finally, the storm has to last at least 3 hours to count as a blizzard. It takes special climate conditions to produce these effects. These conditions include pressure and moisture.

Falling snow does not necessarily mean a blizzard is happening.

Snow does not have to fall for a storm to be a blizzard. High winds can cause a blizzard even when no snow is falling. This weather event is called a ground blizzard. Ground blizzards occur when winds whip up snow already on the ground. Although no snow is falling, the lack of visibility means that the storm is a blizzard.

Other blizzards can produce heavy snow. A snowfall rate of more than 1 foot (0.3 m) of snow per hour is considered heavy. At that rate, the snow really adds up. A very heavy snowstorm can produce more than 10 feet (3 m) of snow before the storm ends.

UNDER PRESSURE

The atmosphere is the air surrounding Earth. It might be surprising to learn that air

Weather systems can be seen in Earth's atmosphere from space.

has weight. But Earth's gravity acts on air the same way it does on everything else. The force applied by air upon Earth's surface is called atmospheric pressure. Changes in atmospheric pressure can affect the weather. These changes in pressure, combined with wind, move weather systems around the planet.

Air rises when it gets warm. This movement creates lower air pressure near Earth's surface. Low pressure usually creates cool, cloudy weather. Areas of very low pressure can create powerful storms, including blizzards. Air sinks when it gets cold. This results in areas of high pressure.

Clouds can move through the sky at more than 100 miles per hour (161 kmh).

High pressure usually brings clear skies and fair weather.

Where areas of low pressure and high pressure meet, **fronts** form. Violent storms can occur when fronts collide. The collision of high- and low-pressure areas also creates strong winds.

KEEPING IT COOL

Blizzards need cold air and moisture to form. Warm air usually has lots of moisture. Wind pulls cold air from Earth's poles. At the same time, wind blows warm air away from the equator. When cold air meets warm air, the warm air rises above the cold air. This movement cools the warm air, causing clouds to form. If the air is cold enough, the moisture in the clouds becomes snow.

The air near the ground also needs to be cold for a blizzard to form. If the air near the ground is too warm, **precipitation** will fall as rain. Snow falls when the air near the ground is below freezing. But it can also fall in slightly warmer temperatures if the layer of warm air is thin.

WHEN AND WHERE DO BLIZZARDS OCCUR?

Blizzards occur in most of the northern parts of the world. There, the weather is cold enough to create snow. Northern Europe, Russia, and Canada often

Mountain climbers and skiers must take care to stay safe during blizzards.

get blizzards. In the United States, blizzards commonly occur in the northeastern and north-central regions. North Dakota, South Dakota, and Nebraska have the highest frequency of blizzards. **Meteorologist** Paul Trambley said, "The Dakotas and Nebraska region can expect around two blizzards a year. During active years, several may occur."[2] Blizzards strike northeastern states from Maine to New Jersey.

However, blizzards can also strike the southeast United States. They also sometimes occur at higher elevations in the west. Even tropical countries can experience blizzards on their mountains. At high elevations, temperatures are low and conditions are quite different from those close to the ground.

In the northeastern United States, many blizzards are caused by nor'easters. These are storms that travel up the East Coast. They gather lots of moisture from the ocean. As the storms move north, they encounter freezing temperatures. This produces heavy snow and blizzard conditions. Nor'easters can occur at any time of year. But they are most common between September and April. The Great Blizzard of 1888, which

The Lake Effect

Cold regions near lakes can experience something called lake-effect snow. This occurs when cold air moves across a large body of water. The air picks up a lot of moisture. This can lead to lots of snowfall along the shore. Lake-effect snow is common around the Great Lakes in the United States.

TEN MAJOR BLIZZARDS IN US HISTORY

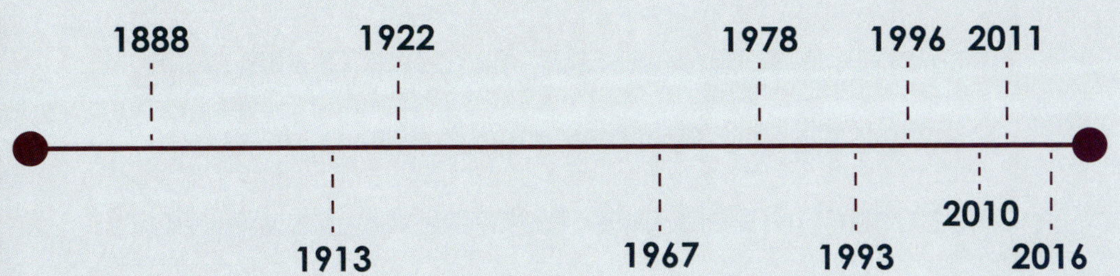

1888: The Blizzard of '88 (Northeast)

1913: The White Hurricane (Great Lakes Region)

1922: The Knickerbocker Storm (South and Mid-Atlantic)

1967: The Chicago Blizzard (Midwest)

1978: Storm Larry (Northeast)

1993: Storm of the Century (East)

1996: Blizzard of 1996 (Northeast)

2010: Snowmageddon (Atlantic Coast)

2011: Groundhog Day Blizzard (Midwest)

2016: Winter Storm Jonas (Midwest and Northeast)

Blizzards occur every year in the United States, but only some become historically powerful storms.

happened in March, was caused by a nor'easter.

A SURPRISE STORM

Blizzards are most common in the winter. But they can also occur in the autumn or spring. On April 6, 1982, a surprise blizzard hit the northeastern United States.

Drivers surprised by an unexpected blizzard may need
to pull over and wait until visibility improves.

It produced up to 18 inches (46 cm) of snow. The storm had started as rain. But the rain turned to snow as temperatures dropped below freezing.

The next day, much of New York and New Jersey was buried in snow. Baseball games were canceled. Airports and train lines shut down. Many travelers were stranded. The storm then moved north into New England. It caused car accidents and power outages. It was the worst April blizzard ever recorded in the region. New York meteorologist Ed Yandrich said, "A blizzard is unheard of here in the month of April." Vermont Highway Department dispatcher Ray Burke joked, "The robins are over their heads in snow."[3]

CHAPTER TWO

THE EFFECTS OF BLIZZARDS

Blizzards can have a big effect on people and their communities. These storms can bring injury, death, and disaster. They can cause huge financial damage as well.

Blizzards put people in danger in many different ways. One of the biggest dangers comes from the cold. People who are outside during a blizzard can suffer

People caught in blizzards must take care to cover exposed skin.

MON THRU FRI
Broadway
FDN

from frostbite. Frostbite happens when the skin freezes.

Severe frostbite can kill body tissue. If body parts are severely frostbitten, they may have to be removed. That is what happened to Joey White. White was trapped outside during a blizzard in Buffalo, New York, in late December 2022. A neighbor named Sha'Kyra Aughtry heard him calling for help and brought him inside. Aughtry said, "He was [. . .] frozen, his pants were frozen, his shoes were frozen."[4]

Aughtry knew she had to get White to the hospital. But the snow made it almost impossible. She called 911, but rescuers could not reach her. Finally, Aughtry posted for help on social media. Someone saw her post and drove to her house in a truck.

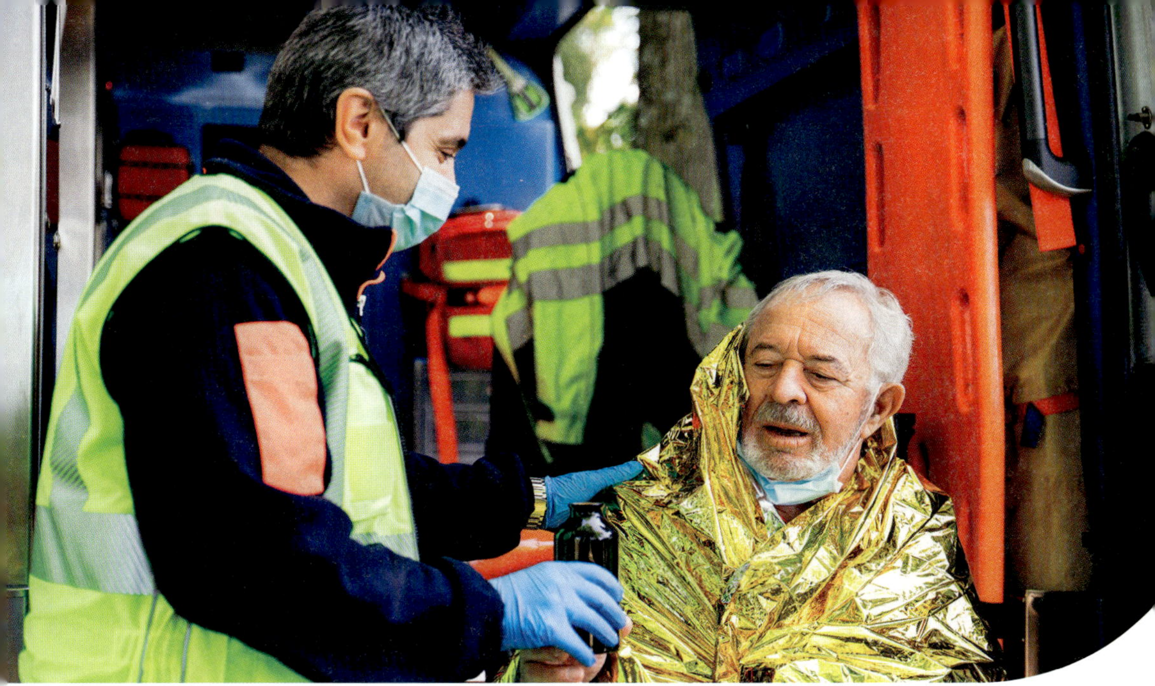

They took White to the hospital. Doctors tried to save White's fingers. But they were too badly damaged by the cold. White lost almost all of his fingers. The efforts of Aughtry, the truck driver, and the doctors kept him from losing his life.

DANGEROUS COLD

Another danger comes from hypothermia. Hypothermia occurs when a person's body

Barns help keep farm animals warm during blizzards.

temperature falls to dangerously low levels. This happens when the body loses heat faster than it can warm itself. People caught in a blizzard might develop hypothermia.

Hypothermia affects the brain. A person who is freezing to death might not even know it.

Animals are also at risk in blizzards. Wild animals can be in danger if they cannot find shelter. Pets left outside can freeze to death. The same thing can happen to livestock in unprotected fields. In October 2013, 100,000 cattle froze to death in a blizzard in South Dakota. Because the storm struck early in the season, the cattle's heavier winter coats had not grown in yet.

The Deadliest Storm

One of the deadliest blizzards ever recorded struck Iran in February 1972. The storm lasted for a week and killed more than 4,000 people. Up to 26 feet (8 m) of snow fell, burying entire villages.

SNOW DANGERS

Blizzards in mountain areas can cause avalanches. An avalanche occurs when a large mass of snow moves quickly down a mountain. The rushing snow buries everything in its path. One of the deadliest avalanches in history buried the town of Yungay, Peru, in 1970. More than 18,000 people died.

People also face danger digging out from a blizzard. People have suffered heart attacks while trying to shovel heavy snow after a storm. Barry Franklin of the American Heart Association said, "Shoveling snow is a very strenuous activity, made even more so by the impact that cold temperatures have on your body."[5] He explained that cold temperatures raise people's blood pressure and constrict the coronary arteries. This makes heart problems, such as heart attacks, more likely.

Snow that has begun to melt is heavier and more difficult to shovel than dry snow.

TROUBLE FOR TRAVELERS

Blizzards create lots of transportation troubles. It can be difficult to drive cars and buses on snowy or icy roads. People may lose control of their cars and get into accidents. Drivers may not be able to see through the snow. Cars may become stuck in heavy snow, stranding people in the storm.

Trains also face difficulties moving along snow-covered tracks. Severe cold can cause ice to form on tracks or freeze overhead power lines. Storm conditions and poor visibility make air travel dangerous. In such conditions, flights are often delayed or canceled.

Travel troubles during blizzards do not just affect people's ability to get around.

Ice on roads is called black ice because it can be hard to see.

Storms can cause delays in the transportation of goods. Freight trains, planes, and trucks may be unable to travel. Businesses may have trouble getting the supplies they need to fill orders.

Ships are also affected by poor visibility, high winds, and rough waves during blizzards. In November 1913, a storm called the White Hurricane struck the Great Lakes.

High winds over water can produce ship-rocking waves.

Twelve ships sank, and 250 sailors died. Modern ships are better at avoiding storms. Captains receive up-to-date weather information. Some ships have computer systems that change routes based on the weather. Large cargo ships have stability systems that keep the ship upright.

HEAVY SNOW, BIG TROUBLE

Blizzards cause damage to buildings. Heavy snow **accumulation** on a roof can cause the roof to collapse. In 1922, a blizzard dumped more than 2 feet (0.6 m) of snow on Washington, DC. The roof of the Knickerbocker Theater collapsed. Ninety-eight people died and 133 more were injured. President Warren G. Harding

A thick layer of snow on a roof can weigh thousands of pounds.

called the event a "terrible tragedy, staged in the midst of the great storm."[6] That storm became known as the Knickerbocker Storm after the disaster.

In 2018, the roof over the pool at a motel in Ashwaubenon, Wisconsin, collapsed after a blizzard. Fortunately, staff and guests

Trees that fall onto roads are usually removed by state or local authorities.

heard the roof cracking and were able to escape. The motel collapse was just one of several roof collapses in the area after the storm.

High wind and heavy snow can also bring down power lines and trees. Many areas experience power outages during storms. Outages leave residents without heat, light, water, or internet service. These incidents can also affect communication with emergency services. Fallen branches can block roads and fall on houses and cars.

AFTER THE STORM

A blizzard's damage doesn't end when the snow stops falling. There is a risk of flooding after a blizzard. Heavy snowmelt can cause

Spring flooding may happen when rain falls on melting snow.

rivers to overflow. Flooding can also happen
when snow melts at a faster rate than the
ground can absorb it. In December 2022,
a strong blizzard hit Buffalo, New York.
Around 51 inches (130 cm) of snow fell on
the city. As temperatures began to rise, the
snow melted. This caused fears of flooding.
Officials put out lots of sandbags in order to
direct floodwaters down safe paths.

Ice is another danger after a blizzard.
Snow melts during warm afternoons. This
leaves roads and sidewalks wet. At night,
temperatures may drop below freezing.
The result can be icy and dangerous
road conditions.

CHAPTER THREE

DEFENDING AGAINST BLIZZARDS

A surprise blizzard can be deadly. However, meteorologists have ways to forecast and track blizzards. Their work helps communities stay safe.

Meteorologists have not always had the technology that they use today. In the past, they tracked blizzards by watching cloud and wind patterns. They also looked at how previous storms behaved.

Modern meteorologists use radio towers to predict the weather.

Technology provided a better way to warn people that a blizzard was coming. By the early 1900s, telegraphs and telephones were used to send storm reports across long distances. These reports helped scientists predict and track storms.

Despite these improvements, it was still difficult to warn people about

One of the first spoken weather reports by radio was broadcast in 1921 by a station in Saint Louis, Missouri.

severe weather. Weather offices were
not staffed 24 hours a day. Instead,
bulletins were issued a few times a day
over the radio. They were also printed
in the newspaper. Sometimes updates
were displayed on cards in public places
or delivered by the mail service. Updates
could also be received by telephone. Since
reports were only issued a few times a
day, people were not always aware about
coming storms. This led to many deaths
because people were unprepared for severe
weather ahead of time.

TECHNOLOGY TO THE RESCUE

Today, people have more technology to
track storms. Two of the most important
tools are satellites and radar. The National

Weather Service began using radar in 1942. Radar works by sending out radio waves. These waves bounce off objects, such as precipitation. Then the waves return to the radar system. They show meteorologists a storm's location, size, speed, and direction.

In 1992, the Weather Service began using Next Generation Radar (NEXRAD). This type of radar could show precipitation and wind movements inside a storm. These clues helped forecasters figure out how severe a storm was.

The first weather satellite was launched on April 1, 1960. Satellites are sent into space. They take photos of Earth's surface. In 1966, the United States launched a geostationary satellite. This kind of satellite orbits at the same speed that Earth rotates.

Weather balloons float high into the atmosphere and record information about the weather.

That means it stays above the same spot on the planet. Geostationary satellites can show how weather changes in that place over time. Scientists use images from these satellites to observe weather conditions around the entire planet.

Meteorologists also use computer models to forecast blizzards. Computers gather information about storms and the

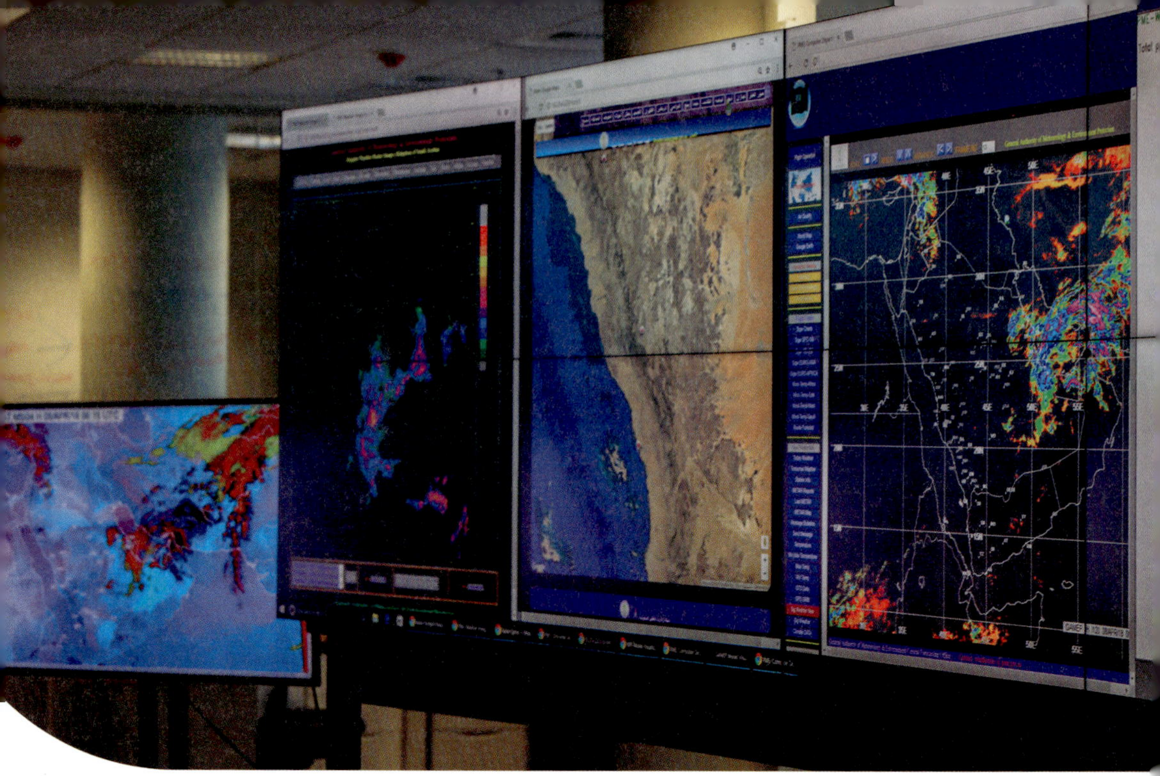

Computer models of weather have become more accurate over time, but they are not always correct.

areas they move over. They use this data to predict where storms will go next and how severe they will be.

However, computer models are not always accurate. Different models can forecast different storm paths. Conditions can change suddenly. Models might not include enough information. For example, in 2000, less than 1 inch (2.5 cm) of snow was predicted to fall on Washington, DC. But the storm dumped more than 1 foot (0.3 m) of snow onto the city. The computer models had failed to consider how thunderstorms over Georgia would affect the weather in Washington, DC.

ALERTING THE PUBLIC

Alerting people to danger is the most important way to save lives during a blizzard. Meteorologists have several ways of getting the word out. The National Weather Service issues storm watches, advisories, and warnings. Local governments and the news media send these messages to the public. These alerts tell people in affected areas when storms might start. They also say how long storms could last. Knowing these things allows people to find shelter. Schools and businesses close early. Transit schedules are adjusted.

Watches, advisories, and warnings are not the same. A blizzard watch means a blizzard might develop. A watch is usually issued about 24 hours before a storm is expected to arrive.

A winter storm or blizzard advisory means that bad weather has started or will arrive soon. Advisories often provide specific details about the storm. They might say how much snow is expected or how long the storm will last.

A warning is more serious than a watch or advisory. A blizzard warning is issued about 12 hours before a storm. It means

Signs over highways alert drivers who may be heading toward severe weather.

the storm is expected to last for at least 3 hours.

However, forecasting blizzards is not an exact science. Big storms can fail to happen. Or they can take a different path than expected. On the other hand, what was forecasted to be a minor storm can develop into a severe blizzard. Unfortunately, false alarms can make people unwilling to prepare for a storm.

PREPARING AND STAYING SAFE

There is a lot to do when a blizzard is approaching. Highway departments make sure their plows are ready to go. They prepare for storms by applying salt to roads. Salt lowers the freezing temperature of water and helps keep snow from sticking as quickly. Many communities have started spraying roads with salt water before a storm. Once the storm hits, plow

Naming Winter Storms

The National Weather Service officially names tropical storms and hurricanes. However, it does not name blizzards or winter storms. In 2012, the Weather Channel began unofficially naming winter storms in order to make talking about them easier. One strong winter storm in 2021 was named Uri.

Some meteorologists give regular weather reports on the news.

operators work around the clock to keep roads clear.

Families and individuals also need to make plans before a blizzard hits. Being prepared is key to staying safe during a blizzard. There are many steps that can be taken before a storm hits.

People should stay inside during a blizzard. Nick Chabarria, spokesperson for the Missouri branch of the American

Automobile Association (AAA), said, "We don't recommend anybody go out unless they absolutely have to during the course of the storm. Let the plows, the road crews, get out there, do their work, make the streets safe again."[7] People have died in blizzards just a few feet from shelter when

Special trucks spray salt over roads using devices called salt spreaders.

they could not see nearby buildings. People should be careful using space heaters or other heat sources during blizzards. Fires or **carbon monoxide** poisoning caused by unsafe heating sources can be deadly. Cars, stoves, or grills are not safe to heat a home.

Every home should have enough water and food to last several days for people and pets. A first-aid kit is also essential. Batteries and flashlights should be available in case the power goes out. All mobile devices should be charged. People should have plenty of warm clothes and blankets. They should stay informed by listening to the radio or checking the internet.

People should be prepared for getting caught in a blizzard while driving.

Some cities operate snowplows in the early morning so that roads are clear for those driving to work.

Cars should carry blizzard supplies. People should keep a container of sand or cat litter in the trunk. These materials can provide **traction** if the car is stuck in the snow. People should also keep a shovel and an ice scraper. Blankets, warm clothes, and boots are important items as well. A car should also have a portable charger, jumper cables, a first-aid kit, food, and water.

After the storm, people should dress warmly and be careful on snowy and icy roadways and sidewalks. They should be careful shoveling heavy snow and take frequent breaks. Blizzards can create magical winter wonderlands as well as dangerous snowscapes. These powerful storms must be treated with respect.

An emergency power generator can be useful when a blizzard knocks out a home's main power.

GLOSSARY

accumulation
an increasing amount of something, such as snow or rain

bulletins
short official news statements

carbon monoxide
a colorless, odorless, and poisonous gas released when substances containing carbon are burned

fronts
leading edges of masses of air

meteorologist
a scientist who studies the weather

precipitation
rain, snow, sleet, or hail

traction
the grip of something on the ground

SOURCE NOTES

INTRODUCTION: A STORM TO REMEMBER

1. Quoted in "'88 Blizzard Blasted Princeton," *The Princeton Recollector*, March 1, 1976, p. 1.

CHAPTER ONE: WHAT ARE BLIZZARDS?

2. Paul Trambley, "Professor Paul Thursday — Where Are Blizzards Most Common?" *Weatherology*, February 14, 2022. https://weatherology.com.

3. Quoted in "April Blizzard Slams New York City," *United Press International*, April 6, 1982. www.upi.com.

CHAPTER TWO: THE EFFECTS OF BLIZZARDS

4. Quoted in "Buffalo Man Saved from Blizzard, Loses Fingers from Frostbite," *WGRZ*, January 23, 2023. www.wgrz.com.

5. Quoted in "A Winter Wonderland Can Turn Deadly with Heart Attacks Brought on by Snow Shoveling," *American Heart Association*, December 20, 2021. https://newsroom.heart.org.

6. Quoted in Joel Wurl, "Winter Takes over the News: The 1922 'Knickerbocker Storm' in Chronicling America," *National Endowment for the Humanities*, January 16, 2016. www.neh.gov.

CHAPTER THREE: DEFENDING AGAINST BLIZZARDS

7. Quoted in Eva Tesafye, "How to Prep for Holiday Driving in the Midst of Arctic Cold and Blizzard Precautions," *Nebraska Public Media*, December 21, 2022. https://nebraskapublicmedia.org.

FOR FURTHER RESEARCH

BOOKS

Monika Davies, *Blizzards*. New York: Enslow Publishing, 2021.

Tammy Gagne, *Hurricanes*. San Diego, CA: BrightPoint Press, 2025.

Charlotte Taylor, *Blustery Blizzards*. New York: Gareth Stevens Publishing, 2023.

INTERNET SOURCES

"Blizzard," *National Geographic*, n.d. https://education.nationalgeographic.org.

Christopher C. Burt, "The Blizzard of 1888: America's Greatest Snow Disaster," *Weather Underground*, March 12, 2020. www.wunderground.com.

"Winter Storm," *FEMA*, n.d. https://community.fema.gov.

WEBSITES

Library of Congress
https://guides.loc.gov/chronicling-america-great
-blizzard-1888/selected-articles

This Library of Congress page contains links to several historical sources about the Great Blizzard of 1888.

National Weather Service
www.weather.gov

The National Weather Service website gives information about past, present, and predicted storms in the United States.

Ready.gov
www.ready.gov

Readers can learn how to protect themselves from winter storms at this website. There are also links to other web pages on related topics.

INDEX

IMAGE CREDITS

Cover: © Zoomik/Shutterstock Images

5: © Lindsay Grace Photography/Shutterstock Images

7: © justkgoomm/Shutterstock Images

8: © Nikolay 007/Shutterstock Images

10: © Andrei Stepanov/Shutterstock Images

13: © Pavel Svoboda Photography/Shutterstock Images

14: © Uzo Borewicz/Shutterstock Images

16: © NASA

17: © s_oleg/Shutterstock Images

19: © Torgado/Shutterstock Images

20: © Michal Martinek/Shutterstock Images

23: © Red Line Editorial

24: © Igumnova Irina/Shutterstock Images

27: © Joe Tabacca/Shutterstock Images

29: © loreanto/Shutterstock Images

30: © Aleksandar Malivuk/Shutterstock Images

33: © Andriy Blokhin/Shutterstock Images

35: © Southern Wind/Shutterstock Images

36: © Nightman1965/Shutterstock Images

37: © NetPix/Shutterstock Images

38: © ND700/Shutterstock Images

40: © fireflite59/Shutterstock Images

43: © Daniel Padavona/Shutterstock Images

44: © Arda_Altay/Shutterstock Images

47: © Edward Haylan/Shutterstock Images

48: © Leo Morgan/Shutterstock Images

51: © F. Armstrong Photography/Shutterstock Images

53: © Meir Chaimowitz/Shutterstock Images

54: © Paolo Bona/Shutterstock Images

56: © Roy Harris/Shutterstock Images

57: © Siarhei Kuranets/Shutterstock Images

Joanne Mattern is the author of many books for young readers. Her favorite topics include history, geography, biographies, and science. She loves sharing information with her readers and helping them discover new things. Mattern lives in New York State with her family.